Forest Babies

Elizabeth Carney

NATIONAL
GEOGRAPHIC

Washington, D.C.

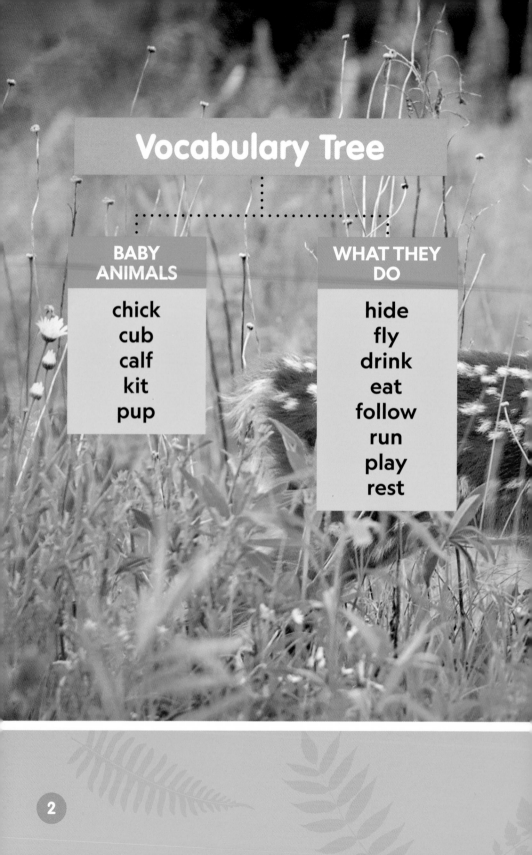

Vocabulary Tree

BABY ANIMALS

chick
cub
calf
kit
pup

WHAT THEY DO

hide
fly
drink
eat
follow
run
play
rest

white-tailed deer

Can you spot the baby?

The forest has everything young animals need to grow.

raccoons

fox snake

A baby snake hatches. Then
it hides under leaves.

A chick stands in its nest.

It flaps its wings to learn to fly.

common
whitethroat

mountain lions

A cub gets a ride
from its mother.

She carries it in her mouth over the forest floor.

A caribou calf drinks milk.

But soon it will eat leaves and moss.

caribou

Cheep, cheep! Hungry chicks open their beaks.

velvet-fronted
nuthatches

**Their mother drops in food
she found in the forest.**

This bear cub snacks on berries. Yum!

black bears

Mother bears show their cubs which plants are safe to eat.

Beavers make their home
out of sticks and mud.

American beavers

A beaver kit follows its
mother out of a pond.

Wolf pups start a game of chase.

gray wolves

They run through the forest meadow.

Squirrel kits play on branches.

American red squirrels

Eurasian red squirrels

These tired kits curl up to rest.

Babies grow, play, eat, and sleep in the forest!

red foxes

YOUR TURN!

Copy these forest babies!

Stick out your tongue like a baby snake.

Stretch like a mountain lion cub.

Howl like a coyote pup.

Turn your head like an owlet.

To Desmond James —E. C.

moose

Published by National Geographic Partners, LLC, Washington, DC 20036.

Designed by Anne LeongSon

The author and publisher gratefully acknowledge the expert content review of this book by Dr. Lucy Spelman, senior lecturer at RISD and founder of Creature Conserve Inc., Providence, Rhode Island, and the literacy review of this book by Kimberly Gillow, principal, Chelsea School District, Michigan.

Photo Credits
Cover, Lubos Chlubny/Adobe Stock; 1, Christopher Loh/Getty Images; 2-3, hkuchera/Adobe Stock; 4, bettys4240/Adobe Stock; 5, Aaron of L.A. Photography/Shutterstock; 6-7, mikelaptev/Adobe Stock; 8-9, Jupiterimages/Getty Images; 10-11, Tatyana Tomsickova/Alamy Stock Photo; 12-13, prin79/Adobe Stock; 14, robertharding/Alamy Stock Photo; 15, Philippe Clement/Nature Picture Library; 16-17, Suzi Eszterhas/Minden Pictures; 18-19, John E Marriott; 20, All Canada Photos/Alamy Stock Photo; 21, Albert Visage/Minden Pictures; 22, Bryant Aardema-bryants wildlife images/Getty Images; 23 (UP LE), Andrew DuBois/Alamy Stock Photo; 23 (UP RT), svehlik/Adobe Stock; 23 (LO LE), Dee Carpenter Originals/Shutterstock; 23 (LO RT), Eric Baccega/Nature Picture Library; 24, Design Pics Inc/Alamy Stock Photo

Library of Congress Cataloging-in-Publication Data
Names: Carney, Elizabeth, 1981- author.
Title: Forest babies / Elizabeth Carney.
Description: Washington, D.C. : National Geographic Kids, 2023. | Series: National geographic readers. Pre-reader | Audience: Ages 3-5 | Audience: Grades K-1
Identifiers: LCCN 2021061247 (print) | LCCN 2021061248 (ebook) | ISBN 9781426373701 (trade paperback) | ISBN 9781426339820 (library binding) | ISBN 9781426374463 (ebook other) | ISBN 9781426374470 (ebook)
Subjects: LCSH: Forest animals--Infancy--Juvenile literature.
Classification: LCC QL112 .C36 2023 (print) | LCC QL112 (ebook) | DDC 591.73--dc23/eng/20220106
LC record available at https://lccn.loc.gov/2021061247
LC ebook record available at https://lccn.loc.gov/2021061248

Printed in the United States of America
23/WOR/1